Reviewing Your
Data Transmission Network

P R D Scott

PUBLISHED BY NCC PUBLICATIONS

British Library Cataloguing in Publication Data

Scott, P. R. D.
  Reviewing your data transmission network.
1. Data transmission systems
2. Computer networks
I. Title
001.64'404   TK5105.5

ISBN 0-85012-326-7

First published in 1983 by:

NCC Publications, The National Computing Centre Limited, Oxford Road, Manchester M1 7ED, England.

Typeset in 11pt Times Roman and printed in England by UPS Blackburn Limited, 76-80 Northgate, Blackburn, Lancashire.

ISBN 0-85012-326-7

Acknowledgements

Much of the material upon which this book is based, together with the appendices, was provided by PACTEL Ltd.

Thanks are also due to individuals from the following organisations who provided guidance at the early planning stages of the project and commented on a preliminary draft of the book:

Christian Salveson Ltd
ICI Pharmaceuticals
Northumbrian Water Authority
Reed International Ltd

The Centre acknowledges with thanks the support of the Computers, Systems and Electronics Requirements Board for the project from which this publication derives.

Contents

1 Introduction

GENERAL

The 1970s were the decade of the computer network. At the beginning of the decade on-line networks were only for the adventurous, but by the end of the decade they were commonplace, and the adventurous had moved on to discover the uncharted territory of 'local area networks'.

This book is about conventional data networks or 'wide area' networks as they are sometimes called in order to differentiate them from their newer counterparts. Essentially the review described is an exercise in how to 'know your network', in terms of how the users view the system, how the network is performing internally, and how much it is costing.

Carrying out a review requires dedicated effort and consumes time and resources. Many organisations will first have to acquire the measurement tools which allow such a review to be undertaken, and so results cannot be made available overnight. The benefits can, however, be considerable. Not only will a review help to identify areas of inefficiency and excessive cost, but it can also provide a firm factual foundation for further development of the network.

It is possible to identify several distinct phases in the development of a network:

— assessment of need;

— design;

— implementation;

— post-implementation review;

— operation;

— enhancement;

— replacement.

It takes one or two years to build a network following the initial design, and most UK data networks are now in the operation/enhancement phase. The introduction of digital leased circuits could, however, have a major impact on today's data networks; the relatively low cost of high-speed digital transmission may encourage redesign and premature replacement of existing networks. Nevertheless, replacement in most cases will be two or three years away, and much can be done in the meantime to improve the efficiency of the present network.

This book describes a review to be carried out during the operational phase of a network's life. It is not a post-implementation review, nor is it a complete network redesign. It is rather an assessment of how the network is currently meeting the demands placed upon it.

In many cases, current demands differ considerably from those envisaged at the design stage. Few networks grow in a completely ordered manner, in strict conformity to a predefined strategy. Haphazard growth in a variety of directions is usually superimposed upon any planned expansion pattern, and this results in sub-optimum solutions. Also, rapid technological progress means that yesterday's state-of-the-art communications hardware may no longer be cost-effective in today's network.

These two facets of changing demand and technological progress provide opportunities for making savings. The aim of the review is to provide a framework through which the network manager can identify these opportunities and take appropriate action.

The post of network manager as such does not exist in many organisations. Typically, a company with a data communications network of any size will have a network controller, responsible for day-to-day network operation, who reports to the operations manager. The one lacks the remit and the other the time to manage the network in the true sense of the word; ie to:

— oversee day-to-day operation;

— plan the future course of the network in terms of enhancement, expansion, modification and replacement;

— evaluate the network's effectiveness;

— control expenditure.

Yet with the growing importance of communications, it is vital that an organisation's communications systems are properly managed. In the absence of forward planning, expedients which solve a short-term problem can become impediments to longer-term development.

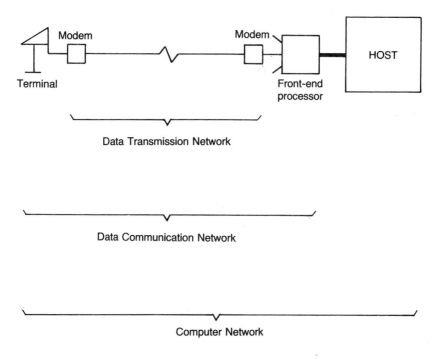

Figure 1.1   Terms and their Usage

It is unfortunate that certain equipment suppliers have coined the term 'network management' to describe their fault diagnosis and control systems. These 'network management' systems are essentially tools to use in solving daily operational problems. True network management, much more a strategic activity looking towards long-term goals and seeking the best route forward, demands both managerial and technical skills.

In summary, reading this book will not on its own make for a better network. A network review requires time to be set aside and probably some expenditure on equipment if real benefit is to be achieved.

DEFINITIONS

Definitions are given, because of the different meanings attached to words like 'network'. Figure 1.1 shows how the following terms are used:

— data transmission network;

— data communications network;

— computer network.

The *data transmission network* refers to lines, modems, concentrators and the like, and is bounded by the digital (V.24) interfaces of the modems at either end of a circuit.

The *data communications network* encompasses all the other hardware and software dedicated to the communications function, including the terminals, front-end processor and any teleprocessing software in the mainframe.

The *computer network* is everything, including the host processor.

Where 'network' on its own is used, the context will normally make clear which type of network is being discussed. This book is basically concerned with the data transmission network, although inevitably it overlaps into other areas on occasions.

2 Quality of Service

INTRODUCTION

This chapter deals with the parameters which indicate the *user's* perception of the quality of service offered by an on-line system. (It does not cover technical performance parameters such as line utilisation which, although of interest to the dp department, have little relevance to the end user. They are dealt with in Chapter 3.)

It is commonly observed that the user's view of a service differs from the provider's view of that same service, and this is one of the main causes of disharmony between user and provider. One reason for this difference in perceived quality is that user and provider employ different criteria against which to judge the service. Measures employed by the service provider are often meaningless to the user, while user measures may appear imprecise and vague to the service provider.

The dp department operating an on-line system finds itself in a dual role as both a *user* of British Telecom's communications services, and a *provider* of service to its own network users. This ought to qualify it to be the ideal service provider but, as many end users can testify, this is not always the case.

Disharmony is best avoided by the dp department, in the context of available resources, agreeing with each user department:

— what parameters should be measured in order to give an objective indication of the quality of service;

— what target figures should be set for each parameter.

13

A fundamental assumption is, of course, that the type of service provided meets the user's requirements.

In an on-line system, the two parameters of terminal response time and availability are the most commonly used indicators of quality of service. They are meaningful to the end user, fairly easily measured by the dp department, and give a good overall picture of system performance. They are not the only parameters which can be measured, but because of their universality they merit detailed discussion. It must be remembered, however, that there are many other aspects of quality of service which do not lend themselves to simple measurement, such as:

— the 'friendliness' of the dialogue;

— the 'friendliness' of the user-support facilities;

— the 'friendliness' of the switch-on/log-on procedure;

— documentation;

— confidentiality;

— fault reporting and clearing procedures.

All these contribute to a user's view of the system.

An organisation needs channels of communication via which user's views on these other aspects of service can be passed back to the dp department. Possible channels include:

— regular dp/user department review meetings;

— user satisfaction surveys;

— visits by dp staff to user sites;

— visits by users to computer centre;

— on-line reporting system;

— 'suggestions box' approach, possibly on-line.

The remainder of this chapter is devoted to a discussion of availability (and the related parameter reliability) and terminal response time. It covers:

— how they are defined;

— how they are measured;

— how they can be improved.

SYSTEM AVAILABILITY

Availability is the proportion of time (typically the proportion of the working day) that a system is capable of performing its designed job for the end user. Availability can be expressed as:

$$\frac{MTBF}{MTTR + MTBF}$$

where MTBF is the Mean Time Between Failure (and is a measure of the reliability of the system) and MTTR is the Mean Time To Repair the failure. MTBF and MTTR are usually measured in hours; only operational time (ie hours when the equipment is supposed to be available) are counted.

The above expression shows up the relationship between reliability and availability. A system can be very reliable (high MTBF) but if on the rare occasion when it does fail it takes a long time to repair, the overall availability will be low. Conversely, an unreliable system can give good availability if the repair time can be made very short.

Consequently it is important for the dp and user departments to agree upon targets for both reliability and availability. This involves deciding:

— the hours of day during which the service should be operational;

— planned down-time for maintenance, where applicable;

— how long the system can be unavailable before business is seriously affected;

— an estimate of likely fault frequency, based on past experience;

— an allowance for the unforeseen.

In some organisations the on-line system is crucial to the minute-by-minute running of the organisation's business, whereas, in others, short periods of down-time can be tolerated. Consequently it is well nigh impossible to lay down norms at which an organisation should aim. Some networks come very close to 100% availability (eg greater than 99.5%), but at this level the cost of achieving even the slightest increase in availability can be considerable.

It is also important to define exactly what is being made available. Is it the availability of:

— the required application programs at all user terminals?

— the required application programs at some proportion of the user terminal population?

— the required application programs at specified user terminals?

— the required application programs on the host machine, with no reference to user terminals?

Ideally the availability of the required applications at each and every user terminal should be known, although practically something less comprehensive than this may have to be accepted. Table 2.1 shows possible reasons for *un*availability, and actions that can be taken.

MEASURING AVAILABILITY

In a multi-component system like a network, it is often easier to calculate availability figures than to measure them directly. Availability of service to the end user (As) depends on the availability of the host (Ah) and the availability of a communications path from the user's terminal to the host (Ac). Hence:

$$As = Ah \times Ac$$

The *host system availability* is readily found from the daily log. The *communications path availability* can be found easily if the network utilises intelligent data communications equipment in conjunction with a network management and control system (see Appendix B). Failing this, it will be necessary to work out the availability of the whole communications path from records of the

Problem	Solutions to consider
Terminal faults	Provide extra terminals, to allow for some terminals being faulty. Supply a pool of spare terminals with facilities for rapid transportation to user sites. Replace terminals by a more reliable model.
Line failures	Provide standby leased lines (preferably alternatively routed), or PSS/PSTN standby facilities. Provide limited processing and storage facilities to remote sites, to permit on-going data collection and validation during line outages, with subsequent batch transmission to the central computer site. Consider alternative network topologies.
Communications hardware faults	Provide switch-over to spare equipment (either manually, via dial-up access from the network control centre, or via integral network control system).
Environmental faults	Provide standby electrical plant and duplicate other essential equipment. Special precautions may be needed in industrial environments to protect equipment from heat, dust, vibration, interference, etc.
Late starts	Management action required.
Unresolved faults	There will always be some faults in this category, but if these become excessive, more sophisticated fault diagnosis procedures are required.

Table 2.1   Reasons for Unavailability

availability of each of the components of the path. Communications path components include:

— terminals;

— communications;

— modems.

Other possible components include:

— multiplexers;

— concentrators;

— controllers;

— sharing units;

— front-end processors.

Clearly, calculation of the communications path availability demands a very good system of documentation to record component down-times. Some compromise may be necessary here to achieve a reasonable measure of availability within the resources of the dp department.

IMPROVING AVAILABILITY

The first step towards improving system availability is to classify the reasons for non-availability. Without such statistics, any efforts towards improving availability stand a good chance of being misdirected. Causes of non-availability include:

— terminal faults;

— line faults;

— communications hardware faults (modems, front-end processors, multiplexers);

— software faults (communications, operating systems, applications programs);

— environmental faults (power, ventilation, etc);

— mainframe hardware faults;

— late starts;

— planned down-time;

— misoperation by users.

'Good fault records will identify the cause and duration of each fault on the network. Effort can then be concentrated where it is likely to yield most benefit.

Communications Equipment

We have already seen that availability can be expressed as:

$$\frac{\text{MTBF}}{\text{MTBF} + \text{MTTR}}$$

where MTBF and MTTR are measured in 'operational' hours. Two approaches to improving that availability are clear:

— increase MTBF;

— reduce MTTR.

Increasing the MTBF lies more within the power of the equipment designer than the user: the user is limited to taking simple precautions such as ensuring that the equipment is neither overloaded nor overheated. Regular meetings with suppliers help to keep equipment performance under review; fault-prone equipment, identified from the user's records, can then be brought to the supplier's attention for corrective action.

Reducing MTTR is a more immediate remedy, and can be readily achieved by providing standby equipment which can be switched into circuit quickly to 'repair' the system. The faulty equipment can then be removed and repaired at leisure. Network control systems allow such switching to be performed remotely from a central control position. Providing extensive standby equipment can be very costly, but is justified when the on-line system is crucial to an organisation's business.

Although lines and equipment do fail spontaneously, there are some occasions when failure is preceded by a deterioration in performance. This can be spotted by carrying out regular error counts on each circuit, but this is tedious to do manually. Network control systems with their comprehensive monitoring facilities will do such checks automatically, and can be programmed to raise an alert when performance deteriorates below a predefined threshold; remedial action can then be taken before outright failure occurs. Often this can be accomplished outside operational hours, and so no loss of availability is incurred. Table B.5 in Appendix B lists the circuit parameters which can be monitored.

Another way of improving availability is to reduce the number of pieces of equipment between the user and the application program, since overall reliability suffers when components are connected in series. However, there are not many situations where this is feasible.

Training

Training represents a less exciting investment than a sophisticated network control system, but it can nevertheless reap benefits in terms of its improved service availability.

User Training

Adequate user training, coupled with a simple pictorial guide on operating the system and dealing with faults, is the best safeguard against user misoperation and spurious fault reports. The training of new staff is important.

Some organisations nominate one person at each site to liaise between users and the dp department. This person may be given the status of a supervisor with additional terminal facilities, or may simply be a coordination point for information passed to and from the dp department. Either way, such people can alleviate some of the load on dp staff. An unfortunate consequence is that the dp department can appear even more remote to end users; however, appointment of a supervisor need not preclude direct contact between end users and the computer centre.

Technical Staff Training

Relations between technical staff at a dp centre and suppliers' maintenance staff depend to some extent on the technical competence of the dp staff. Experience shows that the greater the technical expertise of the person reporting the fault, the faster the response from the maintenance organisation. Training dp staff ensures that suppliers' maintenance people are only called out for genuine faults on their equipment.

RESPONSE TIME

The dp department normally defines response time as 'the time between the user signifying the end of an input message and the first character of the reply being received from the application program'. (Intermediate characters, required by way of acknowledgement or for status purposes, are ignored.) We shall refer to this as *'terminal* response time'.

It is important that response time does not become the sole focus of attention. In many applications the time taken to input the message and to display or print out the reply are equally important. Again, for some users, merely gaining access to a terminal can be a problem. These other aspects are touched upon briefly below before considering terminal response time in detail. Figure 2.1 illustrates the various components of response time.

TERMINAL ACCESS TIME

Terminal access time is the time that elapses between the user deciding to carry out a transaction and his being able to enter the input data on a terminal. Where a user has his own terminal permanently connected to a particular application, access time is minimal. Delays occur when:

— terminals are shared by a number of staff;

— dial-up access is used;

— security procedures are over-elaborate.

The solution in each case is straightforward, although additional expense may be needed. Difficulties in this area are not immediately apparent to the dp department, and some form of user/dp department liaison is vital if problems are to be brought to light.

INPUT TIME

The time taken to input the data to the terminal depends very much on the application and the dialogue design, which are beyond

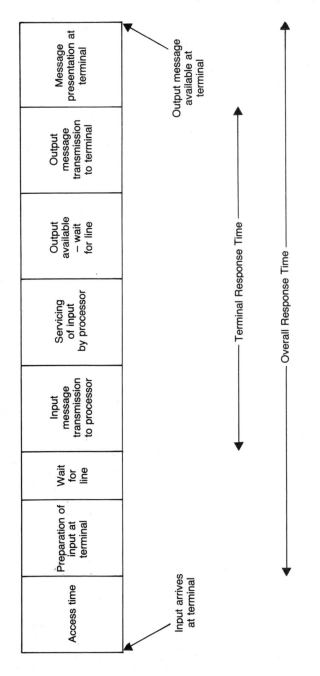

Figure 2.1   Response Time Components

the scope of this book. Several techniques are available for minimising both the quantity of data entered and the quantity of data that is sent to line, and many systems offer varying degrees of 'Help' to suit different categories of user. (The reader is referred to the bibliography for further reading.)

OUTPUT TIME

Message output time at the user terminal is another component of the overall response time (Figure 2.1), and it depends on two things: the rate at which the terminal can display the data and the quantity of data to be displayed. Effective dialogue design is the remedy for problems in this area.

TERMINAL RESPONSE TIME

Terminal response time can be considered as comprising two major components:

— network delays;

— host delays.

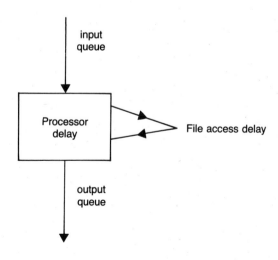

Figure 2.2   Source of Delay in Host Computer

Host delays include input and output queues, and processor and file access delays within the host computer (Figure 2.2).

Normally, file accesses are a major source of delay with processing time being negligible in comparison. However, file access delays, which can be troublesome, are not peculiar to on-line systems, and it is not proposed to discuss them further here. The emphasis in this section is on the delays that occur within the data communications network.

Figure 2.3 shows some of the contributors to network delay. They can be divided into two categories: traffic-dependent delays (which lengthen as the network loading increases) and traffic-independent delays (which remain constant, being a characteristic of the lines and equipment used).

Traffic-Independent Delays

Delays in this category are fixed and unavoidable, but fortunately they are usually relatively short in duration.

Modem Transfer Time

Filters and registers in the modem circuitry introduce delays amounting to a few milliseconds.

Line Turnround Delay

This applies only to half-duplex operation (up to 80ms in UK, 200ms for international circuits).

Propagation Time

The time taken for a signal to travel down a normal circuit in the UK is less than 25ms (45ms for a modem backward channel). Satellite routes add 650ms per up-and-down hop. In circuit switched networks such as the PSTN the exchange introduces no delay; in packet switched networks each exchange introduces a delay. With PSS the average delay across the network is less than 300ms.

Traffic-Dependent Delays

Traffic-dependent delays in the network are caused by polling

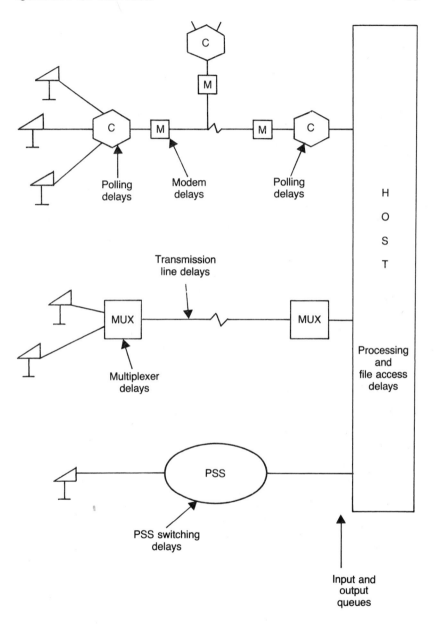

Figure 2.3   Possible Components of Network Delay

routines, and by multiplexers/concentrators and packet switching exchanges which may store data for short periods. These delays are discussed further under the section on 'improving response time'.

MEASUREMENT OF TERMINAL RESPONSE TIME

Terminal response time measurements should be made at a user's terminal, either manually in real-time or recorded by means of a data link analyser for subsequent analysis. However, neither of these techniques lends itself to continuous day-by-day monitoring. In practice the network manager may have to resort to a mixture of measurement and estimate to arrive at terminal response time. Fortunately measurement facilities are normally available for traffic-dependent delays, in the form of:

— hardware and software monitors at the host site;

— statistical output ports on multiplexers and concentrators.

Estimates of traffic-independent delays, which are usually constant for a given terminal, can be found either by experiment or can be approximated from published figures.

Peak-Period Measurements

It may not be necessary to monitor terminal response time throughout the day, every day. Network loading often follows a fairly predictable pattern so that the time of peak loading is known, in which case it is only necessary to measure terminal response time during this peak period. However, on some networks it may be necessary to carry out continuous monitoring for several weeks to determine whether any definite pattern exists. Any such programme of continuous monitoring should, where possible, include foreseeable busy periods (like year end, quarter end or month end) and should avoid weeks where traffic could be expected to be lighter than usual. Average terminal response time should be calculated for every fifteen-minute period throughout each day of the measurement period and then plotted to produce a graph like that of Figure 2.4. This graph is not untypical in having a peak terminal response time several times the average terminal response time. From this graph the busiest one-hour or half-hour period of the day should be selected. If this exercise indicates that

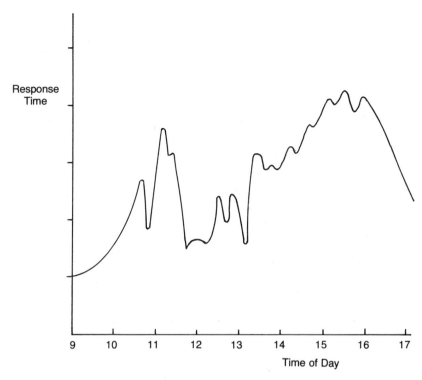

Figure 2.4 Variation in Response Time Throughout the Day

the peak period does occur at a consistent time, response time measurement need only be made around this busy period each day. This technique can significantly reduce the amount of statistical information gathered compared to continuous monitoring. However regular checks should be made to see whether the peak period shifts following the introduction of new applications, changed working conditions or such like.

SETTING TARGETS FOR RESPONSE TIME

A series of response time measurements at a terminal will normally produce a graph of the shape shown in Figure 2.5. A feature of this distribution is its long tail, which indicates the presence of some very long response times. This asymmetry makes it misleading to quote average response times only, and a better solution is to talk

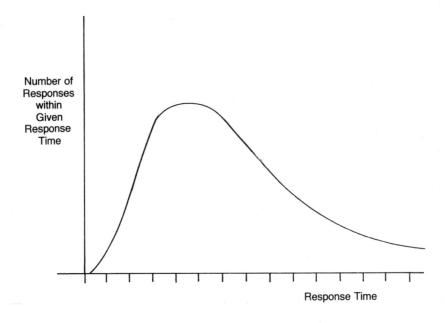

Figure 2.5   Distribution of Response Time Measurements

in terms of the response times achieved by, say, 90% and 95% of all transactions. For example, the target terminal response time might be:

— 90% of all responses to be within 2 seconds;

— 95% of all responses to be within 4 seconds.

In some applications, a two-second response time would be hopelessly slow, whereas in others it would be more adequate. Too fast a response time can lead to careless data input: deliberately slowing down the response time makes the terminal operator look at the screen and thereby check the data that has just been entered. Consistency of performance is important to the user: a consistently fair response is preferred to a sporadically excellent one.

IMPROVING RESPONSE TIME

Network Delays

Two basic ways of reducing delays in the network are:

— to increase the data capacity of the physical circuit;

— to reduce the volume of data being carried on the circuit.

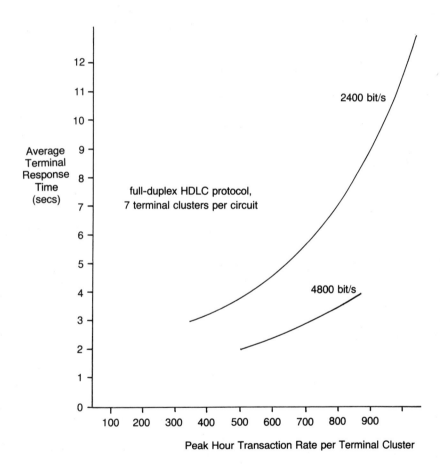

Figure 2.6   Effect of Line Rate on Response Time

Circuit Capacity

Increasing the circuit capacity is most easily achieved by substituting higher-speed modems. This is best done on a trial basis initially, using 'borrowed' equipment, in order to assess its effect. Figure 2.6 shows the improvement in response time that can be achieved by this means.

If half-duplex operation is employed, consideration should be

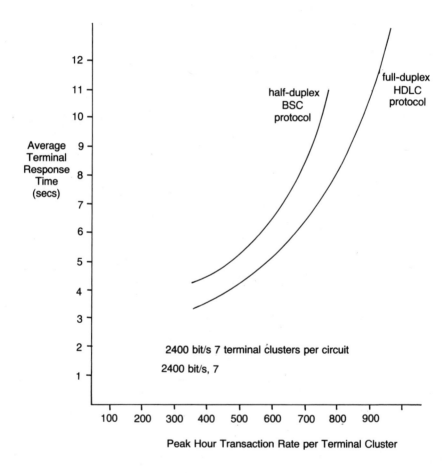

Figure 2.7   Effect of Line Protocol on Response Time

given to changing to full-duplex working. Not only does this avoid
delay in line turnround but it permits more efficient protocols to be
employed, such as HDLC and its derivatives. Figure 2.7 shows the
difference between the full-duplex HDLC protocol and the half-
duplex BSC protocol.

Reducing Data Volumes on the Circuit

Methods of doing this are discussed below. Not all the methods
given will be appropriate to every situation; they should be
regarded more as a menu of possible ideas.

a) Keep messages short by means of:

— efficient dialogue design;

— data compression (only feasible in certain applications);

— encouraging users to operate economically, eg by
using cursor controls instead of repeated spaces.

b) Reduce error rates on the line.

Errors lead to retransmission of data which increases the
number of overhead bits on the circuit. Error rates can be
reduced by:

— using a private circuit instead of a PSTN connection.
The error rate on a good quality private circuit is
approximately 1 in 10^7, compared to 1 in 10^3 on the
PSTN;

— considering a higher quality circuit (if using private
circuits);

— rescheduling data transmission to times outside the
peak periods. The error rate on both private circuits
and PSTN connections tends to increase as telephone
traffic increases, and is at a peak mid-morning and
mid-afternoon. Instances where this is a realistic
option are limited.

c) Reduce the number of terminals on shared circuits (Figure
2.8). On a polled multi-drop line, one rule of thumb is that line

loading should not exceed 40%; above this level of loading, response time increases rapidly.

d)   Alter the polling algorithm so that inactive terminals are polled less frequently. Users could also be encouraged to switch off terminals when not in use.

Other Delays

If a multiplexer or concentrator is causing excessive delay, then it is overloaded and the system needs to be redesigned. One rule of thumb for statistical multiplexers is that the aggregate input traffic

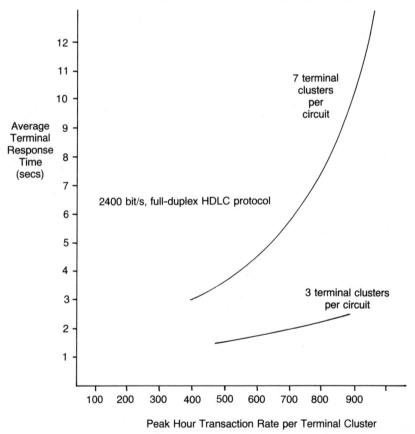

Figure 2.8   Effect of Line Loading on Response Time

should not exceed twice the data rate of the multiplexed circuit. With some multiplexers, certain channels can be given preference over other channels, which allows the delay for some users to be reduced at the expense of increased delay for other users.

Host Delays

Because most processors give preference to output over input, large volumes of output can lead to input queues developing with consequent delays. Bulk output should therefore be avoided at peak periods.

Output Time

The time it takes to display a message at a user's terminal is governed by the data rate of the circuit or by the speed of the terminals in the case of a slow device like a printer. Where a variety of terminal types share a circuit, slow speed devices can be buffered so that the circuit rate is not held down to that of the slowest speed device. Once again dialogue design is an important factor in speeding up interaction with the user.

PRESENTATION OF RESPONSE TIME AND AVAILABILITY MEASUREMENTS

As well as the dp and user departments agreeing upon targets for availability and response time, agreement will also be needed on the frequency and method of presentation of the recorded results. It is common for performance statistics to be averaged on a daily or weekly basis and presented weekly or monthly. Rolling averages (ie calculating the average for the last n days or weeks) can help to iron out short-term fluctuations and highlight longer-term trends. Graphical presentation is easier to assimilate than tables or figures. Figure 2.9 is a example of availability measurements for one network, based on a ten-week rolling average.

SUMMARY

This chapter has presented a methodology for assessing a system performance in an objective way. The methodology involves the user and dp departments:

— agreeing upon quality of service parameters to be monitored;

— defining those parameters closely;

— setting target values for those parameters;

— monitoring the parameters regularly;

— producing summarised reports of quality of service measurements;

— reviewing parameters and targets periodically.

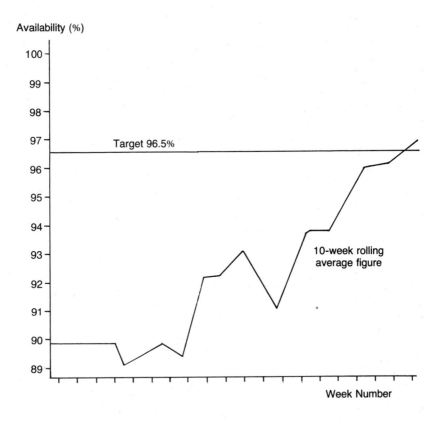

Figure 2.9   Specimen Availability Graph

3 Traffic Analysis

THE NEED FOR TRAFFIC ANALYSIS

If a network is to exist and expand in a cost-effective manner, a complete understanding is required of the traffic flowing on the network. It is necessary to know:

— whence the traffic originates;

— where the traffic is destined;

— what type of traffic it is.

'Traffic' includes all data bits, control bits, acknowledgements, polling signals and any other bits on the line. The units in which traffic is measured may vary from one part of the network to another. They can be bits, characters, frames, blocks, messages or packets.

Unless this traffic is properly analysed, gross inefficiencies can lie hidden within the network. What is more, these inefficiencies tend to be promulgated through successive network enhancements, because they are accepted as the norm.

There are at least two levels at which traffic can be analysed: hardware-independent and hardware-dependent traffic analysis.

HARDWARE-INDEPENDENT TRAFFIC ANALYSIS

At this level, the on-line system can be modelled as a box into which users and application programs send data, and from which they receive data (see Figure 3.1). This is the basis on which the system was originally installed: these traffic flows are the funda-

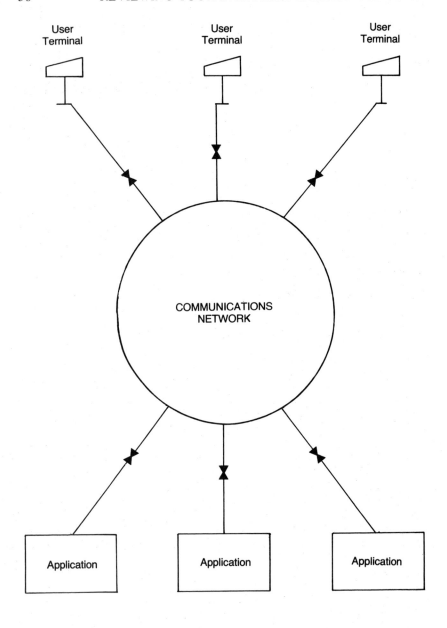

Figure 3.1   Simple Model of an On-Line System

mental information flows for which the network was designed. At this level traffic is totally independent of the communications hardware.

Knowledge of these traffic flows is essential for:

— considering new applications;

— investigating alternative network topologies;

— redesigning the network.

The sort of statistics required at this level are those indicating the pattern of terminal usage; for example, for each terminal at each site:

— activity record;

— hosts accessed;

— applications accessed;

and for each application:

— average input message length;

— average output message length;

— number of transactions/sessions/etc for the day.

HARDWARE-DEPENDENT TRAFFIC ANALYSIS

If every terminal has a dedicated line to a dedicated computer port, as in a simple point-to-point star network, no further traffic analysis is necessary. In most networks, however, there is some degree of sharing of the communications facilities by the different terminals. To make best use of the communications facilities the network manager therefore needs to know the traffic flows at this lower level – the hardware level – where the overheads of line protocol, polling discipline and the like assume importance. At this level, variation in terminal usage patterns tends to be smoothed out and, for most networks, traffic on the network as a whole is likely to exhibit a reasonably predictable pattern, in spite of the fact that it originates from a variety of independent sources.

Traffic analysis at the hardware level allows the network manager to:

— plan for the existing traffic to be handled cost-effectively;

— ensure that capacity exists for growth of existing applications;

— assess the impact of any projected new applications.

Note: New applications may not increase the network loading if they only give rise to traffic outside the busy peak period. However, they may require considerable expansion of the transmission (and computing) facilities, if they coincide with the existing busy period. The other effect a new application can have is to shift the busy period of the network as a whole to a different time.

Traffic details required at the hardware level are line and node utilisation. Line utilisation includes:

— average block size;

— average throughput;

— overheads due to negative polls;

— retransmissions;

— line protocol;

— polling protocol, etc.

Node utilisation (a node being a junction point within the data transmission network at which there is contention, eg a concentrator) includes:

— queue lengths;

— traffic in;

— traffic out;

— transit time.

These parameters will vary throughout the day as user activity varies, but, as already discussed under response time, many networks exhibit a fairly predictable pattern of traffic day-by-day.

Both peak and average utilisation statistics should be recorded so that the network manager can keep track of the ratio between them. A data network is normally dimensioned to meet peak demand, which can be several times higher than average demand.

If the peak demand becomes very much higher than average, however, the network manager may decide that it is uneconomic to cater for peak traffic and instead only dimension the network for, say, 90% of peak traffic demand. Users would then experience a severe deterioration in performance at peak periods. Whether or not this approach is feasible will depend on the business environment and on the purpose of its on-line system.

Alternatively peak demand must be reduced. In a charge-back situation, charges can be increased at peak periods to discourage demand. In other situations, applications which are not time sensitive can be barred from peak periods.

METHODS OF TRAFFIC MEASUREMENT

Traffic levels in a network can be measured using techniques and tools that are either an integral part of the network or attached on a temporary basis for the duration of the measurement. Traffic statistics are available from a variety of sources, including:

— teleprocessing monitors/front-end processor software;

— hardware monitors;

— network control systems;

— remote concentrators/statistical multiplexers;

— software at terminal controllers.

The choice of a particular approach obviously depends on what measurement techniques are already provided in the network (if any) and the cost of bringing in additional measurement resources in relation to the potential benefit.

Traffic statistics produced from these various sources are likely to be in an idiosyncratic format, requiring considerable 'massaging' before they can be made useful. This difficulty may be alleviated in the future as genuine network management systems become available.

Front-end Processors and TP Monitors

Basic front-end processors, as distinct from hardware controllers and multiplexers, commonly incorporate a range of software driven monitoring aids. These aids are normally able to provide reports appropriate to the communications functions being performed, including:

— total data throughput;

— total message throughput;

— traffic at each network location;

— transmission and forwarding delays;

— error statistics;

— polls;

— transmission requests;

— acknowledgements;

— interrupts;

— time outs.

These reports are normally closely geared to the particular supplier's equipment and will be severely limited in scope in the case of most front-end processors. A fairly common approach therefore is to add a teleprocessing monitor: a good tp monitor can greatly enhance the range and detail of reporting, and will be able to provide traffic breakdowns on a per-application basis.

A customising routine which allows the data gathered to be reduced to a meaningful and concise summary can be invaluable to prevent being overcome by the sheer volume of statistics.

Software monitoring at the central site is all that is needed for some networks, but if there is any form of remote concentration in the network, central site monitoring on its own cannot provide a detailed picture of traffic on the network and additional measurements are required using one of the other techniques listed below.

Hardware Traffic Monitors

As an alternative to a software monitor residing in the front-end processor, hardware monitors are available which monitor traffic on incoming ports to the front-end processor by 'bridging' the modem-to-FEP connection. Permanently connected monitors may have a multiple port capability, and provide total traffic observation. A simpler solution is the use of a portable line monitor or data link analyser designed for problem diagnosis as well as for monitoring.

The advantage of the hardware approach is that it decouples traffic monitoring from the network itself – an inherently more robust situation than using the front-end processor to monitor traffic. However the hardware monitor with its limited processing power is unlikely to provide sufficient analysis for the network manager and subsequent processing of the monitor's output will normally be called for.

Network Control Systems

Appendix B gives details of network management and control systems. Most 'network management' systems are in fact network control systems, in that they are geared towards maintaining the network on a tactical basis to meet short-term and immediate needs, rather than being geared towards the strategic approach of meeting long- and medium-term needs. Thus the statistical information available from such systems is essentially related to the functions of fault control, maintenance and performance monitoring. However, with further analysis and interpretation, this raw data can be converted into meaningful management information.

The minicomputer used in 'coupled' network control systems may have some facilities for processing data but this is likely to be limited. The packages offered by the mainframe suppliers are more comprehensive, although they are likely to be specific to that supplier's network architecture approach.

Remote Concentrators and Multiplexers

The range of traffic statistics available from remote concentrators and time division multiplexers varies widely from supplier to supplier and technology to technology. Sophisticated statistical multiplexers provide highly detailed statistics relating to every aspect of system operation, including:

— terminal port utilisation;

— network port utilisation;

— node utilisation;

— threshold values for detecting excessive error rate or excessive processor loading;

— administration/billing statistics where applicable.

At the other extreme, traditional time division multiplexers provide only minimal information relating to traffic such as port utilisation (%) or total data throughput. The style of statistical reporting will be supplier dependent and fairly rigid in most circumstances.

In special cases (turnkey networks) programmable minicomputers may be employed as remote concentrators providing the opportunity for extending the range of traffic measurements available to meet specialised reporting needs.

Terminal Software and Cluster Controller

Traffic monitoring is normally not a standard feature of cluster controllers and terminal devices, although the level of capability can sometimes be increased by developing terminal firmware or programming the controller. The following range of information may then be available:

— number of terminal polls;

— number of acknowledgements;

— number of timeouts;

— number of messages received;

— number of messages sent;

— number of retransmissions;

— average terminal response time.

SOFTWARE PACKAGES

It can be seen that the collection of traffic statistics is still not as straightforward as it should be, given the importance of these figures for planning cost effective networks. Raw statistical traffic data collection, through using any of the previous approaches, needs to be analysed further, using suitable network performance/optimisation/design software packages, to produce report summaries such as:

— graph plots of response time versus loading in part of the network;

— variations in message rates across the network and at certain periods;

— excessive traffic in parts of the network at certain periods;

— insufficient line capacity;

— utilisation of specific components in the network.

Network modelling/simulation packages are available for purchase, but may be expensive. Alternatively, such packages can be run on the user's behalf by either the mainframe supplier or an independent consultancy, but in this case the user may not have access to all the results of the simulation.

IMPROVING NETWORK UTILISATION

High network utilisation makes for economic operation but can reduce flexibility and inhibit growth. Network utilisation can be considered at the same two levels to which we referred at the beginning of this chapter. At the one level – the global network level – utilisation can be improved by encouraging off-peak use of the system. The aim is to flatten out the peaks and troughs of network usage throughout the day to produce as uniform a pattern as possible. This is an administrative rather than a technical change, and as has already been noted, may not be possible in some organisations where network traffic is the result of external stimuli. The other level is the level of hardware, where the utilisation of individual lines and equipment can be examined. Throughput is perhaps a better term to use here. However, improving throughput must always remain subservient to maintaining targets for the user-perceived performance measures discussed in Chapter 2 – especially response time. Methods of improving throughput include:

— putting more intelligence in the transmission network, eg replacing TDM and FDM multiplexers by statistical multi-plexers, or using down-line polling controllers rather than polling from the central site;

— improving line utilisation, eg high quality lines reduce over-heads caused by line errors;

— using more efficient line protocols. HDLC and other full-duplex bit-oriented protocols are more efficient than the traditional binary synchronous protocols.

It is vital that network utilisation is not increased to such an extent that there is no room for growth. Any system needs spare capacity to cater for growth.

NETWORK GROWTH

Assessing Future Requirements

Growth is likely to occur in:

— the number of terminals;

— the traffic per terminal;

— the applications being run.

Growth in existing applications can be forecast from previous years' figures. Once an application is established, its use tends to grow in an exponential fashion, following the traditional S curve. Extrapolating previous years' demands often suffices for assessing future growth. If traffic is plotted on the log axis of a log-linear graph, with time on the linear axis, then straight line growth is observed and extrapolation is easy.

Forecasting the effect of new applications is more problematical. If the company has a five-year plan for data processing laying down a strategy for the spread of computerisation through the organisa-tion, then the network manager is in a position to assess future traffic based on this plan. In the absence of a corporate plan, the dp department will nevertheless need to forecast user growth. An annual survey in which visits are made to each user department by

a systems analyst equipped with a structured survey form is one
approach* . Typical facts required from user departments are:

— new applications planned, including number of new terminals required;

— new terminals for existing applications;

— number of staff involved in existing on-line system;

— number of staff working on systems yet to be computerised;

— number of staff to be recruited.

Apart from known future requirements some allowance should
be made for unexpected developments, and here the experience of
the analyst and the network manager is needed.

In practice, it may be difficult to gain any clear picture of future
demand from user departments, and the network manager will
often be left to forecast future requirements from his own experience.

Given a forecast of future growth the network manager can
devise a strategy for network expansion. This will entail earmark-
ing part of the capacity of the present system for future growth, and
providing a migration path from the existing configuration, so that
all changes to the network are made in a manner that is consistent
with the long-term plan. Ideally the network will have been
designed in a modular fashion to facilitate expansion.

The proportion of system capacity that is reserved for future
growth will be governed, among other things, by the time it takes to
expand the network. This time includes:

— technical planning of network expansion by the dp department;

— lead time for new lines and equipment;

— installation, acceptance and testing of new equipment;

— contingency allowance.

---

* W. R. Nugent in *Advances in Data Communications Management,*
  Volume 1, Heyden Press, 1980.

We shall call this total time reaction time. This is the time it would take from a request being received from a user for a new service until the necessary facilities were provided to that user, were the dp department to start from scratch. If the reaction time is say two years, then capacity for at least two years growth should be built into the network.

This will ensure that the dp department can meet user demands immediately from spare capacity when that demand actually does occur. Meanwhile new equipment will need to be ordered to ensure that there continues to be spare capacity in the network. The manager of a large network is usually at an advantage here, because there is more scope for improvising solutions to unexpected problems.

SUMMARY

Cost-effective networking demands knowledge of the traffic being carried by the network, ie the traffic:

— to and from each site;

— on each line;

— at each network node;

— at the computer centre;

— to and from each terminal;

— to and from each application program.

This chapter has described the tools and methods required to enable a comprehensive traffic analysis to be undertaken.

Armed with this information, the network manager is in a position to:

— identify inefficiencies within the network;

— optimise network design and performance;

— plan more confidently for expansion of the network.

4 Costs

INTRODUCTION

No system can be properly managed without knowledge of where expense is being incurred. All too often in computer systems, this breakdown of costs has been lacking; without it, no effective control can be exercised. This chapter identifies costs associated with a data network and suggests methods of cost reduction.

NETWORK COSTS

Network costs are attributable to:

— communications hardware;

— communications software;

— line rentals;

— public network costs;

— environmental costs;

— staff costs;

— ancillary costs.

The network manager may not have direct responsibility for all these costs, but he may need to take them into account when assessing the economics of different alternatives.

A prerequisite to any methodical evaluation of network costs is an inventory of the lines and equipment that comprise the network, recording, at least, supplier details and costs for each network component. There is a good case for putting this inventory on-line; it facilitates updating and can be a boon in evaluating possible alternatives when contemplating changes to the network. Table 4.1 is a list of some of the information that could be held in such a file*. Transmission circuits and software can be included in this inventory, with appropriate modification to the headings.

As with any on-line file, procedures need to be drawn up for read and write access to the inventory. Preferably one person will have responsibility for updating the inventory, although there is no harm in permitting widespread 'read' access. The inventory needs to be updated regularly and completely reviewed occasionally, which may involve the coordination of several departments.

Given such an on-line system, consideration can also be given to putting fault records on-line as well, providing the sort of information given in Table 4.2. However, the security implications need to be assessed: if the file is on the network host, a fault in the network could render the file inaccessible to the network controller at the very time when it is needed most; alternatively if the file is on a separate stand-alone processor there may be no standby facilities for that processor. For these reasons, many people prefer to retain a manual fault logging system.

An inventory of this type is suitable for recording details of hardware, software, leased circuit and environmental costs. Combine it with a software package for manipulating data from the inventory, and the result is a powerful tool for classifying current costs and evaluating future expenditure.

Public network costs need a different treatment, since unlike the above costs they are usage dependent. Telephone bills should be analysed to show:

— cost of dial-up access, classified into normal and standby;

— cost of phones or telex terminals used by the network controller for fault reporting, remote switching, etc.

Where applicable PSS costs will also need to be taken into account.

---

* from W. R. Nugent in *Advances in Data Communications Management,* Volume 1, Heyden Press, 1980.

Section 1 – Basic Data

  Equipment type
  Model number
  Description options, etc
  Serial number
  Internal reference number
  Supplier reference number
  Location
  Date installed

Section 2 – Supplier Details

  Manufacturer
  Supplier
  Supplier contact point
  Date received
  Maintenance authority
  Contract type
  Maintenance contact point.

Section 3 – Financial Details

  Capital cost
  Running cost, including maintenance
  Purchase order number and date
  Lease or rental costs and duration

Section 4 – Environmental

  Electrical power requirements (volts, amps,
  number of phases, consumption)
  Heat dissipation
  Cooling requirements
  Physical dimensions
  Weight

Table 4.1   Hardware Inventory

Section 1 – Maintenance company

Telephone/telex number for fault reporting
Name of contact
Type of maintenance cover

Section 2 – Network details

Circuit identification
Computer port identification
Terminal identification
Standby facilities
Site telephone number and contact person

Section 3 – Equipment details

Date purchased
Date first installed
Date installed in present location

Section 4 – Fault history

Cause of fault
Date and time fault occurred
Date and time fault reported to maintenance company
Date and time maintenance representative arrived on
    site to repair
Date and time fault repaired
Total time equipment out of service

Table 4.2   Fault Record (for each circuit and equipment item)

Staff costs complete the picture of direct network costs. Staff outside the dp department should not be forgotten, such as users who are asked to help with fault location.

ANCILLARY COSTS

Ancillary costs are those costs, such as documentation and training costs, which are not directly relevant to the day-to-day running of the network, but which nevertheless have a bearing upon it. For example, good documentation and thorough training for both dp and user staff leads to quicker fault diagnosis and thus improved availability. The extent to which such costs are included in any cost exercise is at the discretion of the network manager.

COSTING TECHNIQUES

The network manager should by now have a complete breakdown of network costs. Analysis of these figures may reveal obvious opportunities for cost reduction but other possibilities may need to be evaluated more rigorously. Established costing methods like discounted cash flow (DCF) and present value of annual charges (PVAC) are adequately covered in the literature (see Bibliography). Suffice it to note here that in any cost exercise, it is important to:

— include the total cost of equipment over its entire life. This includes maintenance, accommodation, operating costs and depreciation as well as capital costs;

— ensure that all costs are considered at the same price level. Inflation leads to a face value increase in costs but not an increase in real terms. All costs should be 'normalised' to the same base date. Historical annual inflation rates are available from the Central Statistics Office.

OPPORTUNITIES FOR COST REDUCTION
Investigate Alternative Transmission Facilities

a) PSTN vs leased lines

The relationship between PSTN call charges and leased circuit charges is a complex one, as Figure 4.1 illustrates. Cost relativities between leased circuits and PSTN charges have changed considerably in recent years, and will almost

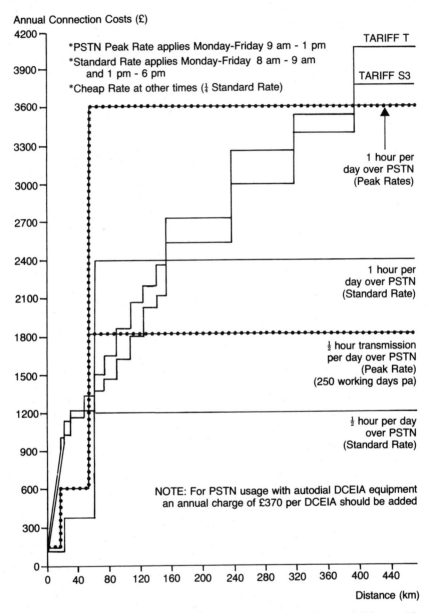

Figure 4.1 Example of Relationship Between Leased Line and PSTN Charges for Data Transmission (based on tariffs current on Feb 1, 1981)

certainly be different now than they were when the network was designed. If the lower quality of PSTN circuits is acceptable, savings may be possible by replacing short distance leased circuits by dial-up connections. Auto-call and auto-answer equipment is available to permit unattended operation. This option may be doubly attractive if an alternative use exists within the organisation for the leased circuits.

b) PSS

PSS call charges are attractive for certain classes of application, but the cost of providing a X.25 port at the computer centre will in many cases outweigh savings in call charges. If, however, an X.25 port can be justified on other grounds – for example, the need to access external databases or to communicate with other organisations – then PSS could prove more economic than leased lines for some intra-company communication.

Additionally, PSS could provide a more satisfactory back-up service than the PSTN, for asynchronous terminals. Appendix A gives more details of PSS.

c) Use lower quality lines

Modern modems can operate satisfactorily on lower quality lines than their predecessors. Changing from tariff T lines to S3 lines can realise savings of 5-10% in line rentals.

Investigate Alternative Topologies

Scope for change in this area is often limited by what the supplier can offer.

a) Loop topology

Statistical multiplexers now offer dual port facilities which allow loop networks to be created, as in Figure 4.2. Line savings can be realised in moving from a star to a loop, but the loop may not offer the same level of security, although it can provide alternative routeing.

b) Mesh topology

A mesh network requires intelligent nodes to handle the

routeing of data and provides a very secure reliable network. All public networks adopt a mesh topology, but it is only justified at present for *large* private networks with several host sites. It is the ideal topology for an organisation seeking a general-purpose communications network. Mesh networks can employ circuit switching or packet switching.

Integrate with Voice

Most organisations which utilise voice and data communications networks have in the past developed, managed and maintained these networks as separate entities. The reasons for this are clear enough, in that the hardware and lines used for voice have not been suitable for high-speed data transmission applications. Technological developments however are changing this situation, and

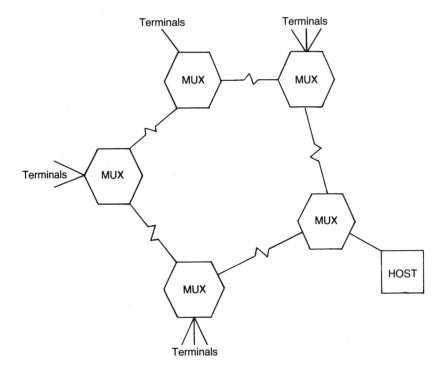

Figure 4.2   Loop Networks Using Statistical Multiplexers

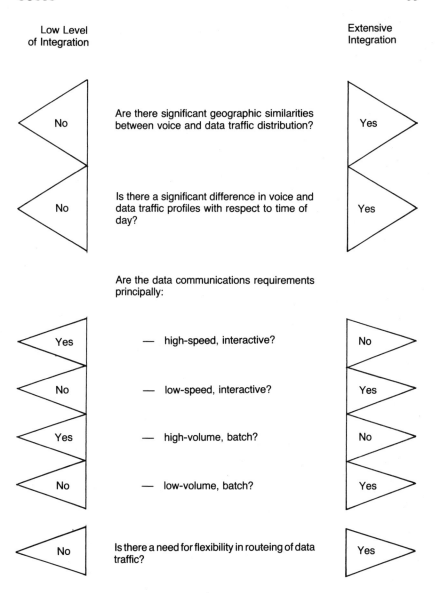

**Figure 4.3   Factors Influencing the Extent of Integration of Voice
and Data Networks**

increasing investment in communication facilities is providing significant impetus to the adoption where possible of common network facilities.

Levels of integration of voice and data communications network facilities can vary from extensive sharing of lines, switching and multiplexing equipment to simple rationalisation of network topology to take maximum economic advantage of tariff structures. The value of extensive integration can be assessed initially by considering the questions contained in Figure 4.3. These questions can only be effectively answered if there is very close cooperation between managers responsible for data and voice communications.

Examples of a low level of integration and extensive integration are shown in Figures 4.4 and 4.5. Typical cost savings of low level integration (Figure 4.4) as against individual circuits for voice and data without grouping, could amount to several thousand pounds per year depending upon geographical locations. The principal benefits of extensive integration (Figure 4.5) lie in cost savings and flexibility. Care is needed in design to ensure that sufficient switching and line capacity is available for both voice and data requirements. This may be difficult or costly to achieve where peak voice and data traffic flows occur simultaneously.

Ideally, large organisations should have one manager with overall responsibility for all aspects of telecommunications in order to promote maximum rationalisation of expensive facilities.

Share Lines and/or Equipment

Multi-dropping is one form of line sharing which is commonly practised, but there are a variety of other devices on the market which allow lines and equipment to be shared, thus reducing costs. Statistical multiplexers can make a significant contribution in this respect.

Appendix C describes the capabilities of such equipment, and covers:

— multiplexers;

— concentrators;

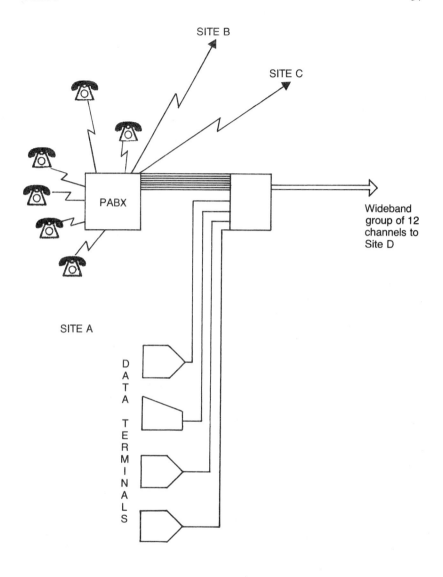

Figure 4.4   Low Level of Integration through Rationalisation of Circuit Provision into one Wideband Group

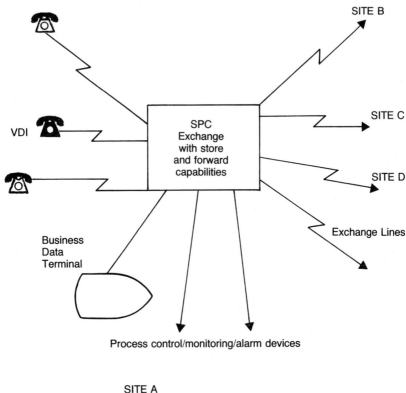

SITE B

SITE C

SPC
Exchange
with store
and forward
capabilities

SITE D

VDI

Exchange Lines

Business
Data
Terminal

Process control/monitoring/alarm devices

SITE A

KEY

VDI   VDI telephone instrument acting as
       a voice and data input device.

**Figure 4.5   Extensive Integration through Sharing of Common
Switching Facilities and Lines**

— line sharing units;

— modem sharing units;

— port sharing units;

— voice/data units;

— X.25 equipment;

— computer exchanges.

STAFF COSTS

Hardware changes can affect both the number of staff involved in network operation, and their skill levels. The trend is towards fuller automation, resulting in fewer staff with lower skill levels.

SUMMARY

Network costs need to be broken down into their component parts in order to obtain a clear picture of expenditure. Only then can alternative solutions be sensibly costed out. An on-line inventory of network components can facilitate cost reviews and exploratory cost studies.

5 Conclusion

GENERAL

Three aspects of a data network have been considered:

— quality of service;

— performance;

— costs.

These represent three pressures on the network manager, all exerting influence in different directions (Figure 5.1). The suggestions for improvement contained in the previous chapters are made from a parochial stand-point, and ignore the impact that an improvement in one area can have in other areas. It is for the network manager to take the wider view and find the *optimum* course of action, consistent with the long-term future of the network. A suggestion has already been put forward that the network manager could benefit from either a mini or microcomputer to help in the management of the network. The primary applications of this machine will be:

— hardware inventory;

— traffic statistics;

— fault procedures;

— cost summaries;

— performance statistics.

The computer could also be used for simple optimisation studies.

Comprehensive network optimisation demands something bigger however. A number of network design tools have been developed and are available for prediction of network performance and network cost optimisation based on stated performance parameters.

Such packages can provide predictions of network behaviour, taking into account:

—   tariff structures;

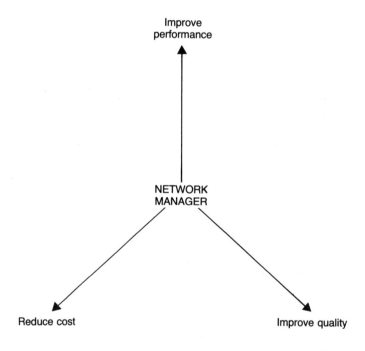

Figure 5.1   Influences on the Network Manager

— protocols;

— workload mix and projected growth and changes in traffic;

— reliability;

— queueing theory;

— routeing algorithm.

There are two principal sources of such network design services:

— equipment suppliers;

— independent consultancies.

Most of the major suppliers of computer hardware utilised in teleprocessing systems make use of such software on behalf of their customers, to assist them in implementing and enhancing their data communications networks. These suppliers see such services as a marketing and system-sizing aid, and are often prepared to use the packages in conjunction with a customer's dp staff. The output from such analysis is usually discussed and shown to the customer but retained by the supplier.

Design tools provided by independent consultancies are usually available either on a licensing arrangement to be used by a client's staff or as part of a wider ranging data processing assignment. Results of analysis are retained by the client.

SECURITY

The increasing importance of data communications, and the transmitting of sensitive data over networks, has led to the need to consider a new requirement in network optimisation. This is the requirement for data security. Data loss or corruption may arise as a result of system or line failure (unintentional) or as a result of unauthorised access to data (intentional).

Protection against Unintentional Data Loss or Corruption

The following techniques will help to minimise the risk:

— use of HDLC-like protocols incorporating cyclic redundancy check code sequences;

— use of buffered devices which retain a copy of transmitted data in buffer until either a given number of frames have been transmitted without a NACK being received from the other end or a positive acknowledgement (ACK) is received for each frame. Frames received in error can then be retransmitted. These two techniques taken together give an undetected error rate of 1 in 10^{21} over a line with bit error rate 1 in 10^6;

— flow control which ensures that no further data is transmitted to a heavily-loaded node until this load has been reduced (mesh networks);

— use of modems incorporating:

    — adaptive, automatic equalisation;

    — line monitoring aids, particularly in respect of signal quality.

Additional facilities can be incorporated, dependent upon the particular network implementation adopted.

System design at a detailed level will also impact the level of security against unintentional data loss or corruption, eg ensuring collection of data from all remote sites at the end of the day by checklists of site numbers and identification of the source of data received at the computer centre.

Whilst every means should be adopted to avoid such loss or corruption of data, for many, intentional access to and tampering with data is the major concern.

Protection against Unauthorised Data Loss or Corruption

Techniques available to reduce this include:

— badge readers and personal identification numbers to protect computer rooms and terminals from unauthorised access;

— passwords to gain access to data, which can operate at file, record or field level;

— data corruption both during transmission and/or where data is stored.

This third area is one in which data communications suppliers are becoming more active. The details of data encryption keys and algorithms and their relationships are not appropriate topics for discussion here. Suffice it to say that:

— encryption may be of the link variety or end-to-end (see Figures 5.2 and 5.3);

— the cost of providing encryption is decreasing at a time when the cost of unauthorised access or corruption is potentially very high and rising;

— every organisation should be assessing:

— the value of the information transmitted within the organisation;

— the likely cost of this information falling into the wrong hands;

— the cost and operational impact of incorporating encryption into an existing network.

The costs and impact consist of:

— encryption device hardware costs;

— additional transmission overheads in terms of:

— delays;

— any additional characters transmitted which will vary depending upon the specific implementation adopted.

Incorporating encryption into an existing network must therefore be embarked upon only after very thorough study.

The only way to ensure comprehensive protection through encryption is to design encryption facilities into the network's protocols. Since this involves considerable and probably unacceptable upheaval in an existing network, the simplest approach to providing significant protection is by means of the encryption box, which is installed at each end of a link or circuit as shown in Figures 5.2 and 5.3. A summary comparison of the two approaches is presented in Table 5.1.

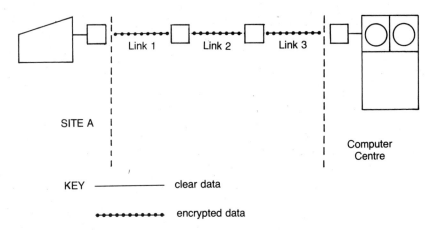

Figure 5.2   Link Encryption (messages are encrypted and decrypted as part of transmission over each link)

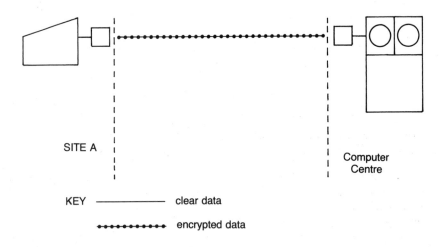

Figure 5.3   End-to-End Encryption (information is encrypted and decrypted only at its destination mode)

Type of Encryption	Characteristics
LINK	Simple key management Encryption and decryption for each link Overheads applied for each link
END-TO-END	Complex key management Clear data only available at end user points Likely to incur less overhead than link encryption

**Table 5.1   Summary Comparison of Link and End-to-End
Encryption**

LONG-TERM PLANS

Security is a good example of a feature that needs to be incorporated into the network at the design stage to be most effective. This reinforces the case already made for a long-term plan for development of the network to be drawn up. Such a plan can considerably ease the task of optimisation since it provides a framework which puts some limits on the number of possible solutions, and thus makes for easier decision-making.

SUMMARY

Network management in the true sense of the word is in its infancy. Efficient, cost-effective networking demands a clear understanding of user needs and a complete knowledge of activity on the network. It is lack of this information that results in the sledgehammer approach to problem solution so often seen on today's networks, with its associated inefficiency and cost.

A review as described here, carried out at intervals throughout the network's life, will enable the network manager to provide a service that continues to provide what users want, in an efficient manner and at an acceptable cost.

Appendix A

Public Switched Networks

INTRODUCTION

Public service offerings are BT's Telex network, the Public Switched Telephone Network (PSTN), and the Packet Switched Service (PSS).

Since these networks are so completely different in terms of technology and capability, and therefore application to data transmission, they will be considered separately in this section. That said, they do have a common attractive feature, in that they are geographically widespread, and therefore provide ready access to all points within the United Kingdom, although in the case of PSS there will occasionally be a need for long access links.

TELEX

The standard Telex service operates at 50 bit/s using the 5-unit code of international alphabet no. 2 (sometimes erroneously called Baudot code). It can be employed for data transmission using 7- or 8-bit codes, with certain restrictions, but it is the low data rate which rules out its use for most data communication applications. It will not be considered further.

PSTN

The Public Switched Telephone Network (PSTN) is used extensively as a vehicle for data transmission. Its use falls into two categories:

— as the normal means of transmission of low-speed, low-volume data (up to 2400 bit/s);

69

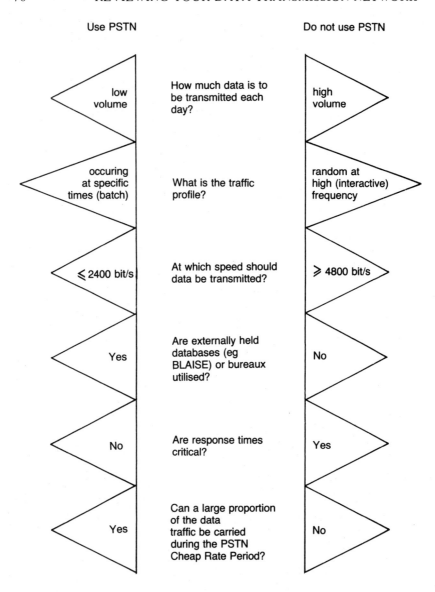

Figure A.1   Factors Influencing the Utilisation of the PSTN as a Normal Means of Data Transmission

— as back-up means of transmission of medium-speed, medium- to high-volume data (up to 9600 bit/s).

PSTN as the Normal Means of Transmission

In considering this issue, it is necessary for the network manager to seek answers to the questions shown in Figure A.1. These answers will indicate whether on balance the use of PSTN as a normal means of data transmission is advantageous.

Note: Attachment of auto-dial equipment and incorporation of auto-answer facilities in modems allows for unattended operation.

PSTN as Back-up Means of Transmission

Star and loop networks are susceptible to failure of a single link or network device resulting in a number of remote locations being disconnected from the computer centre. It has therefore been a common practice to provide back-up transmission capability via dial-up over the PSTN. There are associated costs however, and care is needed if the back-up procedure is to be reliable and efficient, notably:

— back-up should be over direct exchange line (DEL) rather than being routed through a PABX or PMBX switchboard. This is likely to become less important as electronic PABXs are installed, which provide low-noise, high-grade switching and 'do not disturb' indications to the switchboard operators;

— auto-answer facilities can greatly facilitate call set up, and indeed the procedure can be completely automated by incorporation of auto-dial equipment at the computer centre;

— additional costs relate to:

— adaptor units for modems;

— additional DEL rentals;

— provision of dial-up ports on multiplexers, concentrators and front-end processors;

— call charges at the normally applicable rates;

— any special equipment to be purchased/rented (eg auto-dialers);

— any modifications or additional circuitry required on terminal equipment.

Flexible use of the PSTN can improve service availability on star and loop networks significantly (from 98% to 99% is perfectly feasible).

The PSTN is expected to retain this data transmission role with increasing reliability of transmission at 4800 and 9600 bit/s resulting from:

— improved modem technology;

— improvements in quality of connection achieved over the PSTN.

PSS

PSS provides for high quality data transmission at rates up to 48 kbit/s, using packet switching techniques. Being a relatively new service, the role of PSS is only just becoming apparent.

The tariff structure is novel, in that charges are distance-independent, varying only with call duration and volume of data, which tends to favour interactive-type applications where the volume of data is relatively low. However, tariffs are likely to undergo adjustment from time-to-time depending on how traffic on the network builds up.

The two interfaces offered by PSS are an X.25 interface for terminals or computers capable of operating in the packet mode, and an asynchronous interface for simple character terminals. A binary synchronous interface is not offered, but BSC$\longrightarrow$ X.25 protocol converters are available.

Access from PSS to networks in other countries is possible via IPSS (the international packet switching service), and PSS/IPSS has much to offer for international data communications, in terms of quality and cost.

Appendix B

Network Control

DEFINITIONS

The term 'network management and control' refers to a variety of systems, techniques and tools that are used by the network manager/supervisor to improve and maintain network performance levels in response to the needs of network users. Referring to Table B.1, such systems/techniques/tools can be classified in terms of function, ie:

— *Network management:* the maintenance and optimisation of network performance levels on a strategic basis to meet the medium/long-term needs of users. The 'network management' function also encompasses network design and synthesis where this is desirable for optimising and tuning performance;

— *Network control:* the maintenance of network service levels on a tactical basis to meet the short-term and immediate needs of users. This is achieved through performance/status monitoring, and fault diagnosis, in which information is collected remotely and automatically fed back to a central control/computer system for analysis. Furthermore, instructions to reconfigure the network (particularly to by-pass faults) are despatched from the centre to remotely operated switches in the network;

— *Technical control:* similar objectives to network control, although much greater emphasis is placed on the use of test and diagnostic equipment and manually controlled switching/patching for network reconfiguration;

SYSTEM FUNCTION	TASK			
	Reconfiguration	Monitoring and diagnosis	Performance maintenance and optimisation	Network design and synthesis
Network management	✓	✓	✓	✓
Network control	✓	✓	✓	
Technical control	✓	✓		
Test and diagnostic equipment		✓		

Table B.1 Classification

— *Test and diagnostic equipment:* specialised stand-alone devices used to investigate some particular aspect of network performance. In many cases this is a sub-set of 'technical control'.

'Network control' tends to be automated (under the supervision of the network controller) whereas 'technical control' can be labour intensive with heavy reliance on field staff for diagnosing faults and operating switches.

These terms are still the subject of confusion in the marketplace. For example, many suppliers describe their systems as 'network management' or 'network management and control' when 'network control' would be more appropriate.

ALTERNATIVE APPROACHES

The following three main approaches to network control are available, depending on the particular network:

— *Integrated Network Architecture Approach:* network control functions forming an *integral* part of the network's architecture;

— *Coupled Intelligent Network Control System (INCS):* distinct network control system 'loosely coupled' to the network;

— *Manual Test Equipment:* network control performed on a predominantly manual basis – using a range of 'low-level' tools and techniques for monitoring network performance and status.

These three alternatives are illustrated in Figure B.1.

The fully comprehensive approach to network management (in the sense in which it has been used here) is not with us yet, although computer manufacturers and software development companies have been making significant progress in the development of 'intelligent' fault analysis and trend prediction.

INTEGRATED NETWORK ARCHITECTURE APPROACH

The integrated network architecture approach can be the most comprehensive approach although it may be expensive to imple-

(A) Network Architecture

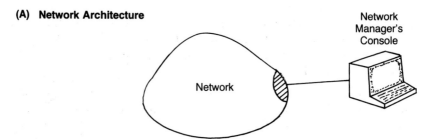

(B) Intelligent Network Control System (INCS)

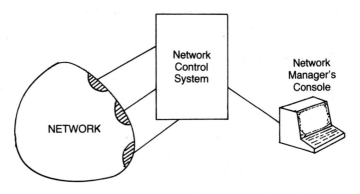

(C) Manual

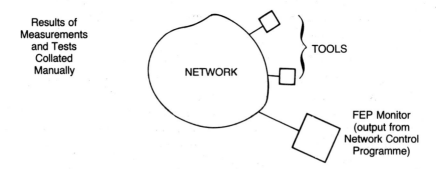

Figure B.1   Alternative Management/Control Configurations

ment and is certainly supplier dependent. The network controller is able to oversee network operation and is provided with a display of network performance and status. He can control log-ons and security checks, and may be given guidance on fault diagnosis in the event of network failure.

Network control facilities in this context are commonly software driven at the level of front-end processor or communications control program within the mainframe. Enhancement of the network control function can be achieved by incorporating a network architecture compatible teleprocessing monitor.

The network architecture approach is becoming far more flexible and adaptable. Network control becomes a built-in function of the supplier's architecture and should therefore be readily able to cater (in time) for any network configuration permitted within the architecture. The approach may, in some cases, be able to provide detailed performance statistics relating to different parts of the network. This avoids excessive dependence on hardware-oriented monitoring and diagnostic information.

In spite of this apparent advantage it is still common practice to supplement the approach with 'technical control' equipment (eg patching panels, T-bar switches) in order that network reconfiguration can be conducted comprehensively. Back-up from individual test equipment is also normally required for investigating local hardware/line problems. Communication between the control centre and devices in the network uses the normal data channel.

INTELLIGENT NETWORK CONTROL SYSTEMS (COUPLED)

Intelligent network control systems are usually supplied by either the modem manufacturer or by an independent supplier. They incorporate a minicomputer at the central site which can access modems and multiplexers in the network in order to collect performance data, and monitor and/or change device status (Figure B.2).

Access can be via PSTN dial-up or private circuits dedicated to this function, but commonly the modem's low-speed secondary

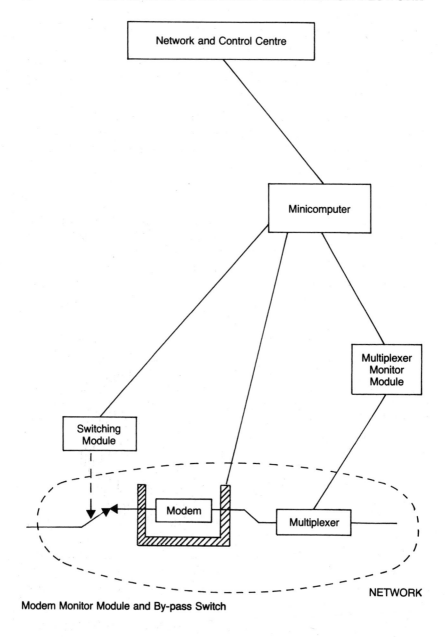

Figure B.2   **Intelligent Network Control System**

data channel is utilised. This channel uses the same physical circuit as the modem's main data channel, but the two are completely independent; monitoring can thus proceed without interruption of data.

SUMMARY

1   Channels in operation
2   Channels out of service
3   Spare port utilisation
4   Spare modem utilisation
5   Patch links available
6   Patch link used
7   Diagnostic bus usage
8   Existing alarms
9   Remote site status

OPERATIONAL

1   New alarms
2   Current interface under investigation
3   Protocol monitor
4   Digital interface status
5   Analogue line under investigation
6   Analogue line level
7   Remote site and line number under control

HISTORIC

1   Operational log with time stamp
2   Line outage with time stamp
3   Alarm arrival with time stamp
4   Alarm acceptance with time stamp
5   Alarm clearance with time stamp

Table B.2   Network Management Capabilities: Information Available

STATUS REPORTS

— remote site status
— digital interface status
— channels operational/
   out of service
— connected/disconnected

— open/closed
— local/remote
— above/below threshold
— under investigation

FAILURE REPORTS

— processor failure
— disk drive failure
— multiplexer failure
— concentrator failure
— device cluster failure
— terminal failure
— link failure

— modem failure
— line failure
— lost data
— time-out
— drop-out
— loop outage
— alert raised

PERFORMANCE REPORTS

— data throughput
— link utilisation
— line utilisation
— buffer utilisation
— queue size
— number of requests for
   retransmission

— number of polls
— number of acknowledgements
— number of interrupts
— error rate
— circuit quality

DIAGNOSTIC REPORTS

— loop back test
— link test
— line test
— modem test
— port test

**Table B.3   Information Provided by Monitoring/Diagnosis
Function**

Such systems permit continuous monitoring of network status and will generate alarms in the event of a fault or when predetermined thresholds are exceeded. The latter facility allows incipient problems to be detected in time to take preventative measures. From the central site, the network controller is able to reconfigure the network and perform remote diagnostic tests. Tables B.2 to B.5 list the type of information potentially available from network control systems.

MANUAL TEST EQUIPMENT

The manual approach is the oldest and least technically sophisticated approach to network control. It is a 'fire-fighting' approach, in that the network controller usually relies on the end user to

---

DIAGNOSTIC RELATED

— select/isolate port
— select/isolate line
— select/isolate modem
— select/isolate monitoring device

ANALOGUE RELATED

— alternate routeing (via standby circuit/link/route)
— dial-up

DIGITAL RELATED

— employ spare port
— employ spare line
— employ spare FEP
— connect patch
— hot standby
— warm standby
— cold standby

Table B.4   Typical Reconfiguration Capabilities

report faults. Fault diagnosis may then require the use of special-ised test equipment by skilled technicians, together with end user co-operation.

The main disadvantages of this approach are that delays can occur before a fault is found and by-passed, and that incipient problems are difficult or impossible to detect because there is no regular performance monitoring.

SELECTING THE APPROACH

The function, topology and size of the network can have an impor-tant bearing on the network approach to be adopted. The relative advantages of the different network control approaches are com-pared in Table B.6 and summarised below:

— the manual test equipment approach employing individual pieces of test equipment, can be very good for hard fault isolation although it is unsuited for monitoring perfor-mance trends on a comprehensive basis;

— the coupled intelligent network control system approach is powerful but is geared very much to the day-to-day (minute-by-minute) control of the network rather than management and trends analysis. It can prove an expensive

```
—  amplitude/frequency response
—  phase/frequency response
—  audio quality
—  harmonic distortion
—  circuit loss
—  amplitude hits
—  signal to noise ratio (white noise)
—  impulse noise
—  phase distortion
—  phase hits
—  phase jitter
```

Table B.5   Measures of Circuit Quality (Line Test)

Approach / Factor influencing decision	Network Architecture	Intelligent Network Control System (INCS)	Specialised, Individual Pieces of Test Equipment
Skill level required	Moderate	Moderate	High
Mainframe supplier	Major suppliers all have offerings, but they may lock users in to one supplier	Can be implemented regardless of mainframe, but cost and implementation usually depends on modems and multiplexers utilised	Can be implemented regardless of mainframe supplier and provides a solution which is very flexible. It can however result in expenditure as high as the INCS approach if it is to provide comprehensive facilities.
Level of sophistication required of system software	High	Low	Low
Costs	These vary with supplier and arise principally from: — software licence fees — demands on hardware, eg memory	Typical costs: – £600 per modem – £350 per communications port	Likely to be high for provision of comprehensive facilities Training of staff may be a significant cost item
Comprehensiveness of operational facilities	Variable	Usually good, but geared very much to control rather than management	Can be very good for hard fault isolation. Unlikely to be so comprehensive in monitoring trends

Table B.6   Comparison of Approaches to Star Network Operation

proposition for smaller networks of less than, say, 100 terminals;

— the integrated network architecture approach is most suitable for complex networks structured around a supplier-dependent network architecture where it may prove particularly cost-effective. Standards of network control tend to vary markedly from one supplier to another.

Some of the issues to be considered in choosing the best approach to network control are listed:

— costs;

— timescales;

— accommodation;

— sophistication/performance:

  — degree of control over the network;

  — degree of automation/self-optimisation;

  — degree of computation/interpretation;

  — speed of response to fault alert;

  — parameters monitored/reported;

— compatibility with future network development;

— dependence on computer/network supplier;

— support available from network management/control system supplier;

— experience elsewhere with similar management/control system;

— software development required;

— availability of suitably qualified/skilled staff;

— impact on existing hardware/software.

Appendix C

Network Hardware

INTRODUCTION

Line costs associated with data communications are increasing at a steady rate.

As the cost of computers and associated terminal equipment falls in real terms, so it would seem that network costs are likely to become a more significant proportion of total dp costs – at least until new technology such as optical fibre cables starts to have a meaningful impact on such costs. It is therefore important that there should be efficient utilisation of data communications lines, consistent with performance requirements.

Modern hardware has a key role to play in this area, and the following sections review the range of equipment available, and the advantages and disadvantages associated with its use under particular circumstances.

MODEMS

Developments in modems over the past few years have given rise to:

— the potential for high-speed data transmission over uncon-ditioned speech circuits;

— enhanced data transmission facilities over the PSTN;

— improved, centralised network monitoring and control facilities, usually utilising a subsidiary low-speed channel for transmission of supervisory signals;

MAXIMUM TRANSMISSION SPEED	TRANSMISSION MEDIUM
1,200 bit/s full-duplex over one line 9,600 bit/s full-duplex using two dial-up connections simultaneously	Public Switched Telephone Network
9,600 bit/s	S3 Unconditioned Leased Line (4 wire)
16,000 bit/s	Tariff T Leased Line (4 wire)
72,000+ bit/s	48kHz Wideband Group

Figure C.1   Transmission Capabilities of Modems

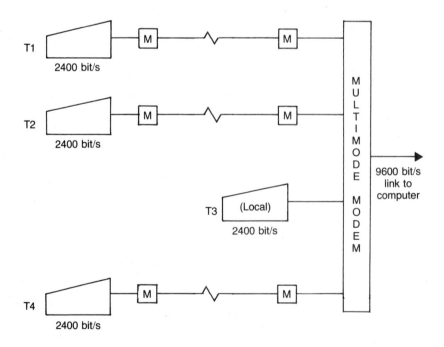

Figure C.2   Typical Use of Multimode Modem

— sharing of modems by a number of terminals by virtue of the multimode modem;

— reduced costs in real terms.

Figure C.1 shows the position regarding transmission capabilities over various transmission media through use of modern modems.

Multimode modems can be used to reduce line and modem costs, effectively combining the functions of modem and multiplexer. A typical application of a multimode modem in this way is illustrated in Figure C.2.

This approach to multiplexing is cheaper than utilising conventional TDM devices, but the penalty lies in lack of flexibility in terms of:

— types of input link supported (synchronous, asynchronous, full-duplex, half-duplex, transmission speeds);

— numbers of input links which can be attached, and in the limitations which can be achieved, in terms of high-speed link utilisation compared with other devices available.

These limitations may result in a large network based on multimode modems being more costly than one based on sophisticated multiplexers and concentrators which provide for highly efficient line utilisation, as hardware cost savings can be more than cancelled out by increased line rentals.

In cases, however, where line utilisation is high and the network is not particularly widespread geographically or supporting a large number of terminals, multimode modems can represent a very cost-effective approach to network design.

MULTIPLEXERS

Three types of multiplexers are briefly considered, namely Frequency Division (FDM), Time Division (TDM) and Statistical (STDM), together with advantages, disadvantages and the circumstances in which each type of device should be utilised.

FD Multiplexers

This technique allows a number of individual terminal lines to share one high-speed data link, by means of sharing the available bandwidth amongst the individual terminals lines. An FDM device acts on analogue signals and is totally transparent. A typical arrangement is illustrated in Figure C.3.

Terminals T1, T2, T3 share the bandwidth of the high-speed link. Downstream line could be multi-dropped rather than point-to-point as depicted here.

The advantages and disadvantages are:

Advantages	Disadvantages
Relatively cheap	Inefficient use of bandwidth
Simple	Inflexible

This device can be used where capital outlay on equipment must be minimal and/or requirements are modest and will remain so.

Conventional TDM

Devices utilising this technique enable a number of terminal lines to share one high-speed data link to a central computer. In this case, signals from each terminal are assigned specific time slots in which to transmit their data, the data from each terminal being interleaved with that from other terminals connected to the multiplexer.

There are two types of TDM – bit multiplexing and character multiplexing – the first is preferred for synchronous working whilst the second is preferred for asynchronous, allowing start and stop bits to be stripped from incoming data before onward transmission along the high-speed data link. Such bit-stripping results in data compression which can give rise to higher efficiency.

These devices allocate transmission bandwidth on a fixed basis, however. Any terminal not transmitting is still allocated a timeslot, giving rise to relatively inefficient utilisation of bandwidth where

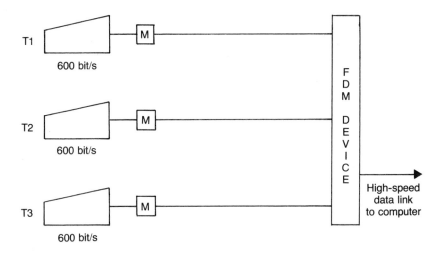

Figure C.3   Typical FDM Configuration

terminal line utilisation is low. Many of the TDM devices offered today can accommodate both synchronous and asynchronous terminals, and can be built up in modular fashion to allow for growth.

Where asynchronous terminals are utilised, the input traffic capacity can be up to three times the output capacity.

Figure C.4 illustrates a typical arrangement using a conventional TDM device. Any downstream line could be multi-dropped, rather than point-to-point as shown.

The advantages and disadvantages of conventional TDM are:

Advantages	Disadvantages
Relatively cheap	Relatively inefficient in an enquiry/response environment
Very reliable in service	

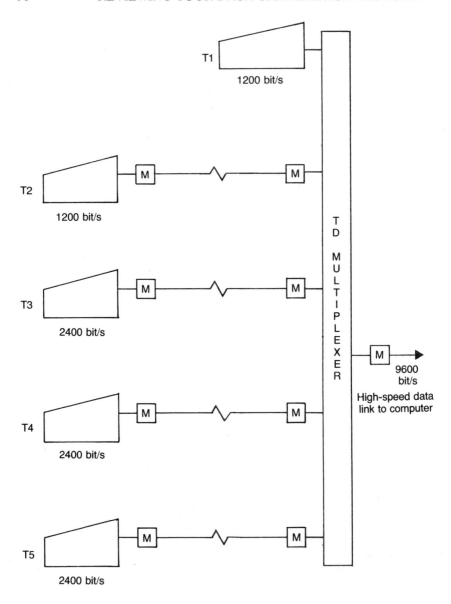

Figure C.4   Typical TDM Configuration

This type of multiplexer should be used where traffic is heavy from and to individual terminals, so that the TDM device is relatively efficient.

Such equipment is particularly appropriate where there are many asynchronous devices transmitting at relatively low speed which are distant from the computer, and where contention between terminals on different lines is regarded as unacceptable. (In the above arrangement, the data communications configuration allows T1, T2, T3, T4 and T5 to have simultaneous access to the computer.)

Statistical Multiplexers

These devices are replacing conventional multiplexers. Their main differences from conventional TDM devices are:

— transmission bandwidth is allocated to terminals according to need at any particular moment in time (dynamic bandwidth allocation);

— they utilise efficient link control type protocols on the high-speed link, in some cases enhanced by sophisticated coding techniques to achieve data compression;

— buffer memory is provided which enables:

 — data to be held for subsequent retransmission in the event of a line break of up to several seconds;

 — data to be held temporarily in the event of very high-density traffic being encountered;

 — other useful data to be held, eg network configuration data, utilisation statistics;

— provision of sophisticated error checking capabilities.

Where asynchronous terminals are used, the input traffic is usually between two and four times the output traffic.

A typical arrangement would be as shown in Figure C.5. Any downstream line could be multi-dropped as opposed to point-to-point as shown.

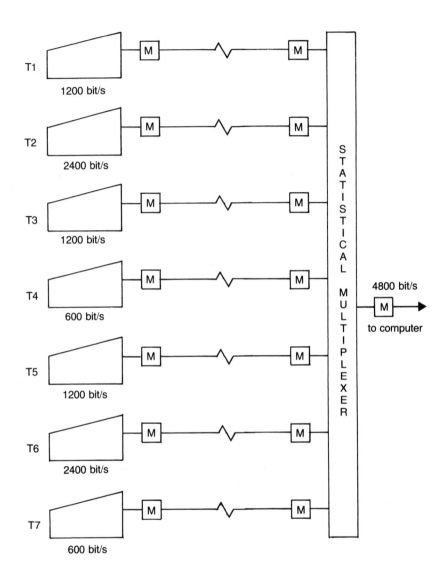

Figure C.5   Typical Statistical Multiplexer Configuration

Advantages	Disadvantages
Very efficient bandwidth utilisation	More expensive than conventional TDM
High reliability	Care is needed in considering traffic from each terminal – high
Highly flexible	traffic will result in buffering of data and possible time-out
Provides error correction	problems in a polled environment

Statistical multiplexers should be used where:

— traffic from each teminal is relatively light;

— flexibility is important in terms of numbers and types of terminals which can be connected;

— line quality may be variable (statistical multiplexers are being used more and more on radio links to protect against the worst effects of fading).

Multiplexers are now being offered which provide additional flexibility by virtue of their being able to transmit data over more than one high-speed data link. Which link is utilised at a particular time can be determined by:

— link status, ie one link is normally utilised to transmit all data and the second serves as a back-up link to be used when the first one fails;

— link loading, ie traffic is divided between the alternative links on a dynamic basis, so that traffic load is equalised, and response times optimised.

Multiplexing devices offering multiple high-speed data link capability provide a migration path from star to mesh networks.

CONCENTRATORS

These devices provide the same functions as multiplexers, but with these added facilities:

— reduction in the number of ports required on central site front-end processors;

— polling of remote terminals and remote end error control;

— the capability of storage and forwarding of data to and from different locations;

— enhanced capabilities for capturing statistical data relating to line utilisation, error rates, etc;

— often software programmable;

— usually capable of protocol conversion, which can permit mixing of different terminal types on one multi-dropped line;

— capable of automatic routeing over a number of high-speed data links.

There are two major penalties associated with acquisition of these additional facilities:

— cost;

— complexity, which usually reflects on availability.

Figure C.6 compares multiplexers and concentrators, a choice which often proves difficult for the network manager, though statistical multiplexers now perform a number of functions previously only associated with concentrators, eg buffering during high load periods.

LINE, MODEM AND PORT SHARING UNITS

Line Sharing Units

A typical arrangement would be as shown in Figure C.7. The line sharing unit (LSU) will only allow one terminal location DTE(1), DTE(2) or DTE(3) to have access to the computer centre at any moment in time. Savings are derived from:

— reduction in the number of long distance data communications links;

— reduction in the number of communications ports required at the computer centre.

Such units can usually be installed in cascade, and should isolate all non-active lines from the computer centre, thus eliminating signals or noise from these lines.

Modem Sharing Units

In addition to cost savings, these units can allow for additional flexibility of equipment to be connected at remote sites.

This can be a valuable asset where some terminals are connected to a terminal control unit, whilst some special purpose devices such

Network Requirement	Multiplexers	Concentrators
Minimum cost	✓	Improving as intelligence costs come down
Maximum flexibility re: routeing, terminal connections, store and forward	Improving, eg capability to handle more than one high-speed link	✓
Fast response	✓	Introduce contention and delay under high load
Ease of implementation	✓	
Ease of operation	✓	
Network control and management data collection	Improving as a result of incorporation of memory	✓

Figure C.6   Comparison of Multiplexers and Concentrators

as alarm units or word processors, which are stand-alone, are also required to be connected into the one common data communications network.

Figure C.8 depicts a typical situation where a modem sharing unit may be fitted.

Port Sharing Units

The device shown in Figure C.7 can also act as a port sharing unit for multiplexers, concentrators and central computer equipment. Savings potential is considerable. The network manager must be sure however that resulting contention will not render response times to users unacceptable.

Note: All equipment which permits sharing of facilities will tend to reduce overall service availability to end users, since a single failure on shared facilities will disconnect a number of users rather than just one.

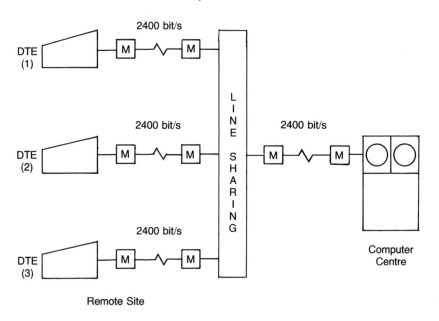

Figure C.7   Use of Line Sharing Unit

VOICE/DATA UNITS

These units allow voice and data communications to be routed over the same lines. Simultaneous, or effectively simultaneous, voice and data transmission is possible using some devices, whilst others only permit voice or data transmission to take place at any particular time.

There are two principal reasons why transmission of voice and data over a common line can be advantageous:

— maximum bandwidth utilisation on expensive leased line facilities, ie economic optimisation;

— network operation often requires voice communication facilities between a network control centre and remote sites.

Units which allow simultaneous, or effectively simultaneous,

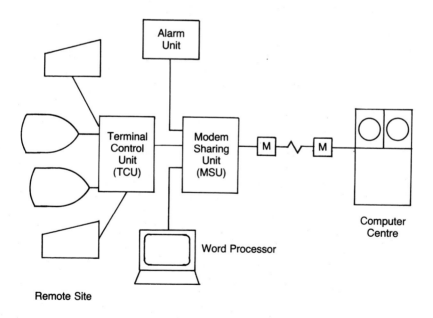

Figure C.8   Use of Modem Sharing Unit

voice and data transmission over one 4-wire voice band line are:

— the speech-plus multiplexer, sometimes referred to as the s+dx panel; available bandwidth for voice and data is fixed;

— the dynamic multiplexer which utilises the pauses in conversations and fills available bandwidth with data.

The essential characteristics of each type of device and situations in which it should be considered are shown in Figures C.9 - C.12.

Terminals may be transmitting and receiving data at speeds of up to 300 bit/s typically.

Where to Use s+dx Equipment

The checklist of Figure C.10 indicates the factors influencing use of s+dx equipment. Note that the system is not compatible with certain speech signalling systems used between PABXs.

More powerful dynamic multiplexing equipment utilising digital technology is available for attachment to high-cost leased lines, eg international circuits, ensuring maximum utilisation of such lines.

Care is needed when assessing the role for such equipment in a private network, and experiences of some users suggest that field trials should be conducted before becoming committed to adoption of such techniques.

There are likely to be restrictions on the transmission mode of

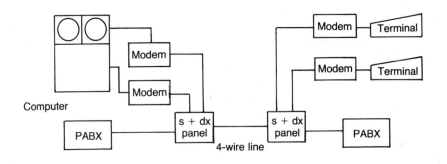

Figure C.9   Typical Network Configuration Utilising s+dx Panels

attached data terminals and computer equipment, eg block transmission on request.

The economic benefits can be very high, however, in those organisations which generate a lot of international voice, data and electronic mail type traffic.

A typical arrangement involving dynamic multiplexing is illustrated in Figure C.11.

The checklist of Figure C.12 indicates the factors influencing use of dynamic multiplexing equipment.

Finally, on the subject of voice/data units it should be mentioned that voice adaptors are offered by most modem suppliers to be used in conjunction with their modems on leased circuits. The central unit allows for speech facilities over a number of lines by

Use s+dx equipment	Factor	Do not use s+dx equipment
Yes	Data volumes between the sites are low	No
Yes	Response time is *not* critical (ie 5-10 seconds acceptable for interactive working)	No
Yes	Voice traffic plus data merit high-quality leased line rental on economic grounds (S3 grade)	No

Figure C.10   Factors Influencing the Use of s+dx Equipment

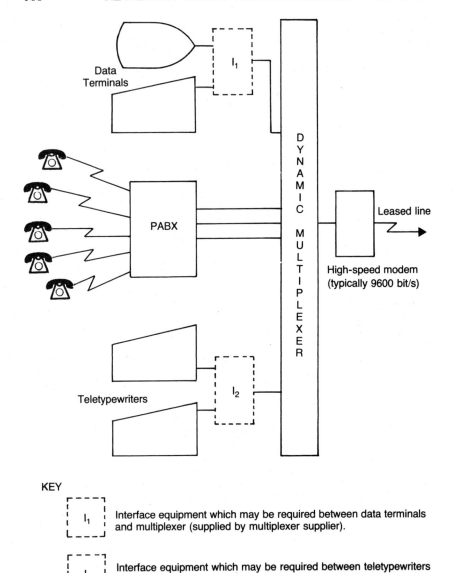

Figure C.11 Typical Network Configuration Utilising Dynamic Multiplexer Equipment

Use Dynamic Multiplexer	Factor	Do not use Dynamic Multiplexer
Yes	High volume of data requiring transmission	No, public network likely to be more cost effective, eg IPSS, Telenet
Yes	International circuit(s) justified for volumes of voice, data, facsimile, etc	No
Yes	Large numbers of high-quality leased circuits required between specific locations within the UK	No
Yes	Unreliable or unavailable public network services between locations in different countries	No

Figure C.12   Factors Influencing the Use of Dynamic Voice Plus Data Multiplexing Equipment

pushbutton selection. Calls can also be initiated from remote sites to the network control centre. In addition to these voice conversation facilities, such units may provide a first-line network control facility, since it is possible to incorporate line monitoring facilities with such units.

SWITCHING DEVICES

Switching devices range from the simple mechanical switches to data switching exchanges.

Mechanical Switches

Figure C.13 shows the arrangements commonly found. Switches of this type can be activated locally by hand or remotely via a telephone circuit, and are used to by-pass faulty equipment.

Intelligent Switches

Intelligent switches cater for a large number of input and output ports (possibly several hundred) and can connect any input port to any output port under program control. Port interfaces may be V.24, 4-wire analogue lines, or asynchronous current loops for local terminals. Such switches can provide contention, with different terminals having different priorities of access, remote dial-in, security passwords, usage statistics, broadcast facilities and the like. Intelligent switches find application in the computer centre to provide contention and/or flexibility, and also out in the network as a remotely controllable switching node.

EQUIPMENT FOR INTERFACING TO X.25 NETWORKS

Although the range of equipment offering X.25 interfaces is growing, many terminals wishing to access either private or public packet switched networks based on X.25 are likely to be asynchronous character terminals or utilise a BSC type protocol (3270 type devices).

For attachment of these types of terminals to X.25 networks, there is a requirement for special interfacing units (PADs – Packet Assembler/Disassemblers), for asynchronous character terminals and the so-called 'black boxes' for BSC type terminals.

Usage of the British Telecom Packet Switched Services

Access by asynchronous terminals is via PADs located at the Packet Switching Exchange. These PADs provide functions based on Telenet software, and there are a number of parameters which can be set by the terminal user. Setting up these parameters can ensure that highly cost-effective utilisation of PSS is achieved, as they determine:

— whether characters are echoed back to the terminals;

— under what conditions data awaiting transmission in the PAD is packeted for transmission over the network – charges are heavily dependent upon the number of packets transmitted;

— how specific signals such as 'break' are to be handled, eg discard output, do nothing.

Any organisation planning to make use of PSS should consider these PAD parameters closely before commencing operation, and produce appropriate documentation for users, which must also take into account the characteristics of the host being accessed.

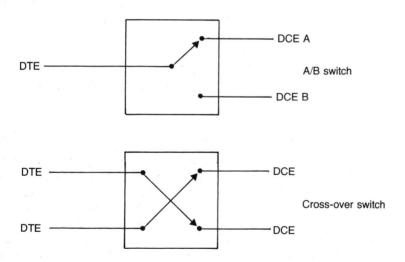

Figure C.13   Mechanical Switches for V.24 Interchange Circuits

In addition to the PADs provided by British Telecom, it is also possible to utilise proprietary approved PAD equipment installed on the user's premises.

The principal benefits of using proprietary PADs are:

— provision of multiple logical channels at host sites;

— provision of X.25 circuit up to the user's premises, with resultant improvements in data integrity.

Access to PSS by synchronous terminals utilising BSC type protocols requires the user to purchase a proprietary 'black box' from an approved supplier.

BSC type protocols show differences between each other as implemented by different suppliers, and for this reason any decision to utilise PSS in a character synchronous terminal environment *must* be preceded by thorough investigation of the following issues:

— availability of terminal interfacing facilities as standard within approved 'black boxes';

— the extent to which attachment of different end-terminal types may require:

  — separate 'black boxes';

  — customising of 'black boxes';

  — BT approval of an, as yet, unapproved device;

  — additional expenditure;

  — additional time and effort to effect migration to PSS.

The need for PADs, protocol converters, etc, in packet switched networks calls for thorough investigation before a decision to proceed is taken. Investment in current hardware and software is likely to be high, and the extent to which this must be modified to permit effective packet switched operation will mean that, for some organisations, use of public or private packet switched facilities may have to await terminal, front-end processor or computer renewal.

Development of Private Packet Switched Facilities

Similar consideration will apply in many cases to those associated with the use of PSS. Additional concerns will be:

— interfacing asynchronous terminals; PADs utilised should conform to CCITT recommendation X.3, X.28 and X.29;

— ability to interface with the Public Network may be useful for back-up purposes. Development of private facilities should, as far as is practicable, track PSS development so as to enable flexible interchange between public and private network facilities.

An implementation of private packet switching networks which was not compatible with PSS would need to show very significant economic or operational benefits to be worthwhile.

Appendix D

Bibliography

Topic	Book
Dialogue Design	*Designing Systems for People,* Damodaran et al, NCC Publications, 1980.
Costing Techniques	*Discounted Cash Flow,* M G Wright, McGraw Hill.
	The Capital Budgeting Decision, Bierman and Smidt, MacMillan.
Data Network Management	*Advances in Data Communications Management,* T A Rullo (Editor), Heyden, 1980.
General	*Handbook of Data Communications* (2nd edition), NCC Publications, 1982.
	Systems Analysis for Data Transmission, James Martin, Prentice Hall, 1972.
	Fault Diagnosis in Data Communications, P J Down, NCC Publications, 1977.

Computer Networks and their Protocols, Davies, Barber et al, Wiley, 1979.

Planning for Data Communications, Bingham and Davies, MacMillan, 1977.

Index